GEOLOGICAL SURVEY OF CANADA
MEMOIR 422

PRECAMBRIAN GEOLOGY OF THE DALY BAY AREA, DISTRICT OF KEEWATIN

T.M. Gordon

1988

Energy, Mines and Resources Canada
Énergie, Mines et Ressources Canada

Available in Canada through

authorized bookstore agents and other bookstores

or by mail from

Canadian Government Publishing Centre
Supply and Services Canada
Ottawa, Canada K1A 0S9

and from

Geological Survey of Canada offices:

601 Booth Street
Ottawa, Canada K1A 0E8

3303-33rd Street N.W.,
Calgary, Alberta T2L 2A7

A deposit copy of this publication is also available
for reference in public libraries across Canada

Cat. No. M46-422E Canada: $10.00
ISBN 0-660-12613-3 Other countries: $12.00

Price subject to change without notice

Cover

Mafic gneiss in Boundary Shear Zone of Daly Bay Complex near Whale Point (T.M. Gordon, ISPG 2611-1)

Critical Reader

F.C. Taylor

Original manuscript received: 1986-06-02
Final version approved for publication: 1986-06-22

Preface

This report documents the metamorphic and structural history of rocks that were initially deposited during the Archean, buried to depths of at least 15 km, and thrust into higher crustal levels during the Aphebian. They form part of a 500 km long belt of Archean middle and lower crust now exposed along Chesterfield Inlet in the District of Keewatin.

Studies such as this, and others elsewhere in the Canadian Shield, are leading to an improved understanding of the architecture and behaviour of deep continental crust. Rather than acting as rigid and quiescent plates, continents are now known to undergo large scale internal ductile deformation, and to contain faults and shear zones that remain active for billions of years.

The influence of these features on the generation and migration of ore-forming fluids is becoming an increasingly vigorous area of research. Continued careful mapping of the Canadian Shield will provide important constraints for all such studies.

R.A. Price
Assistant Deputy Minister
Geological Survey of Canada

Préface

Ce rapport renseigne sur l'évolution métamorphique et structurale des roches qui ont été initialement déposées durant l'Archéen, enfouies à des profondeurs d'au moins 15 km, puis amenées par des mouvements tectoniques à des niveaux plus élevés de la croûte terrestre au cours de l'Aphébien. Ces roches font partie d'une zone de 500 km de long, composées de la croûte à des niveaux moyens et inférieurs d'âge Archéen, et maintenant exposées le long de l'inlet Chesterfield dans le district de Keewatin.

Les études de ce genre, et d'autres réalisées ailleurs sur le Bouclier canadien, nous permettront de mieux concevoir l'architecture et le comportement de la croûte continentale profonde. On sait maintenant que, loin de se comporter comme des plaques rigides et stables, les continents subissent à grande échelle des déformations à élongations internes, et contiennent des failles et des zones de cisaillement qui restent actives pendant des milliards d'années.

L'influence de ces détails sur la production et la migration des fluides minéralisants, constitue de plus en plus un domaine de recherche active. En poursuivant avec soin la cartographie du Bouclier canadien, on obtiendra de l'information importante pour délimiter le champ de ce type d'étude.

R.A. Price, sous-ministre adjoint,
Commission géologique du Canada

CONTENTS

PRECAMBRIAN GEOLOGY OF THE DALY BAY AREA, DISTRICT OF KEEWATIN

Abstract

The Daly Bay Complex of granulite gneiss, gabbroic anorthosite, and migmatite underlies an area of 2500 km^2 immediately north of the Chesterfield Fault Zone in the northeastern Churchill Province of the Canadian Shield.

The complex consists of metasediments of probable Archean age which have been intruded by gabbro and gabbroic anorthosite. Aphebian granulite facies metamorphism (0.55 GPa, 750°C) and isoclinal folding have effected all of these rocks. Mineral assemblages include quartz — K feldspar — plagioclase — sillimanite — biotite — garnet — cordierite; and quartz — K feldspar — plagioclase — biotite — hornblende — garnet — orthopyroxene — clinopyroxene.

A 1 to 15 km wide, inward dipping, ductile shear zone surrounds the Daly Bay Complex and contains strongly deformed equivalents of rocks in the core. Amphibolite facies metamorphism has overprinted shear zone rocks as well as some rocks within the complex. Greenschist facies mineral assemblages are associated with late northeast-trending faults.

The Daly Bay Complex lies on an east trending regional gravity high that extends 550 km from Baker Lake to Coasts Island. Rocks along this zone include granulite gneiss, gabbro, and anorthosite. Based on observations in the Daly Bay Complex, this zone is interpreted as a belt of late Archean crustal rocks, emplaced during the Early to Middle Aphebian.

Résumé

Le complexe de Daly Bay constitué de gneiss granulitiques d'anorthosites gabbroïques et de migmatites supporte une région de 2500 km^2 juste au nord du champ de faille de Chesterfield dans le nord-est de la province de Churchill du Bouclier canadien.

Le complexe comprend des métasédiments probablement de l'Archéen qui ont été injectés par des gabbros et des anorthosites gabbroïques. Un métamorphisme aphébien à faciès granulitique (0,55 GPa, 750°C) et des plissements isoclinaux ont affecté toutes ces roches. Les assemblages de minéraux sont ainsi répartis: quartz — K-feldspath — plagioclase — sillimanite — biotite — grenat — cordiérite; et quartz — K-feldspath — plagioclase — biotite — hornblende — grenat — orthopyroxène — clinopyroxène.

De 1 à 15 km de large, cette zone de cisaillement ductile, plongeant vers l'intérieur, entoure le complexe de Daly Bay et contient des équivalents rocheux fortement déformés en son centre. Un métamorphisme à faciès amphibolite a repris les roches de la zone de cisaillement tout comme certaines roches à l'intérieur du complexe. Les assemblages de minéraux du faciès à schistes verts sont associés avec les failles orientées récentes du nord-est.

Le complexe de Daly Bay se situe sur un maximum gravimétrique régional de direction général est, qui s'étend sur 550 km du lac Baker à l'île Coats. Les roches que l'on trouve le long de cette zone sont des gneiss granulitiques, des gabbros et des anorthosites. D'après les observations effectuées sur le complexe de Daly Bay, on interprète cette zone comme étant une zone de roches de la croûte de la fin de l'Archéen, mises en place au cours de la phase qui s'étend du début jusqu'à l'Aphébien moyen.

SUMMARY

The Daly Bay Complex of gneiss, gabbro and anorthosite underlies an area of 2500 km^2 on the coast of Hudson Bay between Daly Bay and Roes Welcome Sound in northeastern District of Keewatin.

The map area lies within the Churchill Structural Province of the Canadian Shield. All bedrock is Precambrian, with the latest metamorphism probably occurring during the Aphebian. The oldest rocks are gneiss and mafic intrusive rocks of the Daly Bay Complex, which underlies the south-central two-thirds of the map area. In the central part of the complex, the rocks retain granulite facies mineralogy and textures, but towards the margins of the complex they have been strongly sheared, folded, and retrometamorphosed.

Heywood (1966) defined the Daly Bay Complex as the distinct lithological association of gabbro, anorthosite, generally mafic-rich gneiss, and granulites found in the Daly Bay area. The boundary between these rocks and the surrounding gneiss cannot be precisely delineated as it is a complicated metamorphosed shear zone. For purposes of this report, the Daly Bay Complex is thus defined as those rocks which retain either mineralogical or textural evidence of having undergone granulite facies metamorphism. The resulting boundary corresponds closely to that mapped by Heywood (1967).

The oldest rocks are pyroxene granulite and quartzofeldspathic granulite which have been intruded by gabbroic anorthosite and gabbro. Strongly sheared equivalents of these rocks form an inward dipping Outer Shear Zone to the Daly Bay Complex. Retrograde metamorphism of the sheared rocks has resulted in local development of layered gneiss and migmatite.

The rocks surrounding the Daly Bay Complex are mainly various layered gneisses and migmatite. Several zones of thin layered gneiss, schist, and mylonite occur within the layered gneisses and adjacent to the Daly Bay Complex. Veins, sills, and dykes of light grey to pink weathering granite and granodiorite are particularly common in the layered gneiss at the eastern edge of the map area. Two plutons of coarse grained pyroxene monzonite intrude layered gneisses in the northern part of the map area. They form roughly equant stocks 1-2 km in diameter. Several northwest-trending diabase dykes of the Mackenzie swarm cross the area.

Structures are dominated by features reflecting the emplacement of the granulite facies rocks of the Daly Bay Complex into the surrounding amphibolite facies gneiss. Mesoscopic structures and fabrics show a regional zonation inferred to have resulted from a zonation in deformational regimes. Within the core of the Daly Bay Complex, ductile defor-

SOMMAIRE

Le complexe de Daly Bay, composé de gneiss, gabbro et anorthosite, constitue le sous-sol d'une région de 2 500 km^2 localisé sur la côte de la baie d'Hudson, entre la baie Daly et le détroit de Roes Welcome, dans le nord-est du district de Keewatin.

La feuille cartographiée se situe dans la province structurale de Churchill du Bouclier canadien. Toute la roche en place est précambrienne, et la dernière phase de métamorphisme a probablement eu lieu durant l'Aphébien. Les roches les plus anciennes sont les gneiss et les roches intrusives mafiques du complexe de Daly Bay, qui constituent le sous-sol de la partie sud-centrale des deux tiers de la zone cartographiée. Dans la partie centrale du complexe, les roches conservent la minéralogie et les textures du faciès granulitique, mais vers les bords du complexe, elles ont été fortement cisaillées, plissées et ont subi un rétromorphisme.

Heywood (1966) a défini le complexe de Daly Bay comme l'association lithologique distincte de gabbros, d'anorthosites, de gneiss généralement riches en produits mafiques, et de granulites que l'on rencontre dans la région de la baie Daly. On ne peut définir avec précision la limite entre ces roches et les gneiss environnants, étant donné qu'elle constitue une zone de cisaillement métamorphisée complexe. Dans cette étude, on a donc défini le complexe de Daly Bay comme composé de roches qui conservent les indices minéralogiques ou texturaux d'un métamorphisme allant jusqu'au faciès granulitique. La limite que l'on obtient ainsi correspond à celle représentée sur la carte d'Heywood (1967).

Les roches les plus anciennes sont une granulite à pyroxène et une granulite quartzo-feldspathique qui ont été traversées par une intrusion d'anorthosite gabbroïque et de gabbro. Les équivalents fortement cisaillés de ces roches constituent une zone de cisaillement extérieure au complexe de Daly Bay, et plongeant vers l'intérieur. Le rétromorphisme des roches cisaillées a localement favorisé le développement de gneiss stratifiés et de migmatites.

Les roches entourant le complexe de Daly Bay sont principalement des gneiss diversement stratifiés et des migmatites. Plusieurs zones de gneiss finement stratifiés, de schistes et de mylonites apparaissent à l'intérieur des gneiss stratifiés, à proximité du complexe de Daly Bay. Des veines, filons-couches et dykes de granite et granodiorite gris clair, devenant roses à l'altération, sont particulièrement fréquents dans le gneiss stratifié, sur le rebord est de la zone levée. Deux plutons de monzonite à pyroxène, de granulométrie grossière, traversent des gneiss stratifiés dans la partie du nord du levé. Ils forment des amas intrusifs (stocks) approximativement équidimensionnels, d'un à deux kilomètres de diamètre. Plusieurs dykes de diabase de direction générale nord-ouest, faisant partie de l'essaim de dykes du Mackenzie, traversent la région.

Les structures sont dominées par des détails reflétant la mise en place des roches du faciès granulitique qui font partie du complexe de Daly Bay, dans les gneiss environnants du faciès amphibolitique. Les structures et textures mésoscopiques montrent une zonation régionale, qui résulterait d'une zonation des régimes de déformation. Dans le centre du complexe de Daly mation is suggested by continuity of layering and large scale folds. In places, however, mineral textures are granoblastic, suggesting annealing at granulite grade. At the margins of the complex, conditions were such that planar fabrics and finer grain sizes are preserved.

Outward from the Daly Bay Complex, deformation under amphibolite facies conditions produced bands of thin layered and strongly foliated gneiss alternating with less well foliated rocks. Later brittle deformation produced local cataclastic zones with greenschist facies mineralogy. The interaction of metamorphism and deformation means that some rocks probably passed through all metamorphic and deformational regimes, but retain evidences of only the final stages of their journey.

On a regional scale, structures within the complex define an upright, shallowly southeast-plunging synform with a steep southwest limb and a moderately dipping northeast limb. This trend is at a high angle to the northeast regional trends of this part of the Churchill Province. Major shear zones outline, or are parallel to, the margins of the Daly Bay Complex, indicating that it is allochthonous. All rocks in the complex have undergone ductile deformation. The amount of strain, as reflected in grain size, layer thickness, and development of mylonite textures, increases from the core of the complex to its boundaries.

Rocks in the Daly Bay Complex have been metamorphosed in the granulite facies. Mineral assemblages include quartz — K feldspar — plagioclase — sillimanite — biotite — garnet — cordierite; and quartz — K feldspar — plagioclase — biotite — hornblende — garnet — orthopyroxene — clinopyroxene. Within the complex, there is no recognizable gradient in metamorphic conditions. At the margins of the complex, and near Bernheimer Bay, minerals such as garnet and pyroxene have been partially or completely replaced by hydrous minerals such as biotite, chlorite, and cummingtonite. Near late faults, chorite-epidote-white mica assemblages indicate greenschist facies conditions.

Mineral composition data indicate that rocks of the Daly Bay Complex were metamorphosed at 750 ± 50°C and 0.55 ± 0.1 GPa. They reached these conditions prior to and during the initial stages of their emplacement into the adjacent rocks.

A limited amount of geochronologial data is available for the Daly Bay Complex. A U-Pb analysis of zircon from a two pyroxene enderbitic granulite south of Lake of Islands gave a concordant age of 2067 Ma. This date can thus be interpreted as a minimum age for protolith to the gneisses, and a maximum age for granulite facies metamorphism. A K-Ar date from biotite in a pegmatite cutting the foliation in the Boundary Shear Zone is 1951 ±

Bay, la continuité de la stratification et la présence de grands plis suggèrent une déformation à élongations. Toutefois, par endroits, les minéraux ont une texture granoblastique, qui suggère qu'il y a eu une reprise au niveau du faciès granulitique. Sur les marges du complexe, les conditions ont été telles que les foliations et la granulométrie fines ont été conservées.

En s'éloignant du complexe de Daly Bay, on note que la déformation dans les conditions du faciès amphibolitique a produit des bandes de gneiss finement stratifiés et fortement feuilletés, qui alternent avec des roches moins nettement feuilletées. Une déformation fragile ultérieure a engendré localement des zones cataclastiques caractérisées par une minéralogie du faciès des schistes verts. L'interaction du métamorphisme et de la déformation a probablement eu pour effet de faire passer les roches pour tous les régimes métamorphiques et de déformation, mais celles-ci conservent seulement les traces des dernières étapes de ce passage.

À l'échelle régionale, les structures contenues dans le complexe définissent un synforme dressé à la verticale, plongeant faiblement vers le sud-est et comportant un versant sud-ouest raide, et un versant nord-est de pendage modéré. Sa direction fait un angle prononcé avec les directions régionales nord-est de cette partie de la province de Churchill. De grandes zones de cisaillement délimitent, ou suivent parallèlement, les marges du complexe de Daly Bay, indiquant que celui-ci est allochtone. Toutes les roches du complexe ont subi une déformation à élongations. Le degré de déformation, tel que reflété par la granulométrie, l'épaisseur des couches, et le développement de textures mylonitiques, augmente du centre du complexe à ses limites.

Les roches du complexe de Daly Bay ont subit un métamorphisme dans le faciès granulitique. Les assemblages minéraux sont les suivants; quartz — K-feldspath — plagioclase — sillimanite — biotite — grenat — cordiérite; et quartz — feldspath potassique — plagioclase — biotite — hornblende — grenat — orthopyroxène — clinopyroxène. À l'intérieur du complexe, on ne peut clairement identifier de gradient métamorphique. Sur les marges du complexe, et près de la baie Bernheimer, les minéraux tels que les grenats et les pyroxènes ont été partiellement ou entièrement remplacés par des minéraux hydratés comme la biotite, la chlorite et la cummingtonite. Près des failles tardives, des assemblages à chlorite-épidote-mica blanc indiquent les conditions métamorphiques du faciès des schistes verts.

Les données relatives à la composition minérale indiquent que les roches du complexe de Daly Bay ont été métamorphisées à 750 ± 50°C et 0,55 ± 0,1 GPa. Elles ont atteint ces conditions avant et durant les étapes initiales de leur mise en place dans les roches adjacentes.

On dispose d'une quantité limitée de données géochronologiques relatives au complexe de Daly Bay. Une analyse du rapport U-Pb dans du zircon provenant d'une granulite enderbitique à deux pyroxènes, prélevée au sud de Lake of Islands, a indiqué un âge concordant de 2 067 Ma. On peut donc interpréter cette date comme l'âge minimum du protolithe dont proviennent les gneiss, et l'âge maximum du métamorphisme dans le faciès granulitique. Une datation par la méthode K-Ar de la biotite dans une pegmatite qui recoupe la foliation dans les zones

26 Ma, providing a minimum age for cessation of ductile deformation at the margins of the complex.

My interpretation is that the Daly Bay Complex may contain some rocks initially deposited or intruded in the Archean and that granulite facies metamorphism of these rocks occurred at approximately 2050 Ma. This metamorphism was terminated by tectonic emplacement into higher crustal levels which was complete at about 1950 Ma. Further uplift of the complex and surrounding rocks followed at about 1850 Ma.

The Daly Bay Complex lies on an east trending regional gravity high that extends 550 km from Baker Lake to Coats Island. Rocks along this zone include granulite gneiss, gabbro, and anorthosite. Based on observations in the Daly Bay Complex, this zone is interpreted as a belt of Archean lower and middle crustal rocks, tectonically emplaced during the Aphebian.

de cisaillement limitrophe indique un âge de 1951 ± 26 Ma; ceci nous donne l'âge minimum de la fin de l'étape de déformation à élongations sur les marges du complexe.

Suivant mon interprétation, le complexe de Daly Bay pourrait contenir des roches initialement déposées ou intrusives dans les terrains archéens, et le métamorphisme de ces roches dans le faciès des granulites se serait produit il y a environ 2 050 Ma. Ce métamorphisme a pris fin avec les processus tectoniques qui ont amené le terrain à des niveaux plus élevés de la croûte; ce soulèvement s'est terminé il y a environ 1 950 Ma. Un soulèvement ultérieur du complexe et de ses roches environnantes a eu lieu il y a environ 1 850 Ma.

Le complexe de Daly Bay se situe sur un maximum gravimétrique régional de direction générale est-ouest, qui s'étend de 550 km du lac Baker à l'île Coats. Les roches que l'on trouve le long de cette zone sont des gneiss granulitiques, des gabbros et des anorthosites. D'après les observations effectuées sur le complexe de Daly Bay, on interprète cette zone comme étant une zone de roches de la croûte de l'Archéen inférieur et moyen, mises en place au cours de la phase de l'Aphébien.

INTRODUCTION

The Daly Bay Complex of gneiss, gabbro and anorthosite underlies an area of 2500 km^2 on the coast of Hudson Bay between Daly Bay and Roes Welcome Sound in northeastern District of Keewatin. The area mapped for this report is bounded by latitudes 63°45' and 64°30' N and longitudes 87°45' and 90°00' W, and thus includes the Cape Fullerton map area (NTS 55P), the southern half of Daly Bay map area (NTS 56A), and part of Yellow Bluff map area (NTS 46D). The centre of the area is approximately 130 km northeast of the community of Chesterfield Inlet and 280 km west of Coral Harbour. Figure 1 indicates the location and extent of the map area. A RCMP post at Cape Fullerton was abandoned in 1915, and a fishery at the north end of Daly Bay operated only briefly in the 1960's, leaving the area with no permanent inhabitants.

Chartered aircraft afford access to the area throughout most of the year. Ski-equipped aircraft may be used until mid-June and float operations usually begin in early July. The larger lakes have shallow and rocky shorelines, reducing the effective use of floats. Topography suitable for aircraft with oversized wheels is rare.

The area may be reached by boat from Chesterfield Inlet after breakup. The coastal waters have not been charted in detail and must be navigated with care. Canoes and small boats can be used on the coast and larger lakes but the rivers are generally too shallow to provide useful access to the interior.

Although early explorers made scattered geological observations along the coast of Hudson Bay, Low (1906) was the first to provide lithological descriptions and relative ages of some rocks in the area. The high metamorphic grade of the gneiss and the presence of anorthosite intrusions were discovered during systematic mapping in 1964 by Heywood (1966,1967), who outlined and named the Daly Bay Complex. The purpose of the present work was to map the Daly Bay Complex and environs for publication at a scale of 1:250 000 and to elucidate the petrology and structure of the area.

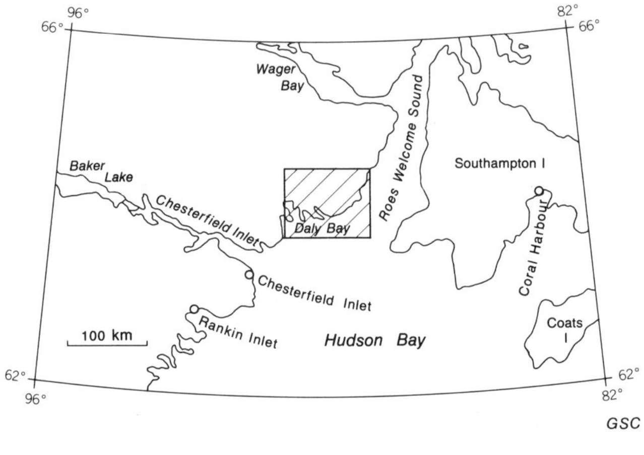

Figure 1. Location of map area. Shaded region is area mapped for this report.

In 1970 the area was reached in mid-July by chartered Peterhead boat. Inflatable rubber boats were used to obtain supplies from Chesterfield Inlet and for coastline traverses. Pace and compass traverses supplemented the shoreline observations. In 1971 a Bell 47 helicopter was available for the first three weeks in June and was used to set out traverse teams and to make spot observations in the northeast part of the area. Fieldwork was completed from fly camps moved by helicopter and Cessna 180.

The Daly Bay Complex lies on the southeastern margin of the Roes Welcome Lowland (Robinson, 1968). Maximum elevations of 250 m are reached at the northern edge of the map area. The upper limit of postglacial marine invasion is approximately 150 m above present sea level and crudely parallels the present coastline. Shoreward from this limit is a 20 km zone of excellent outcrop. Elongate rock ridges follow the structural grain and are separated by steep sided, commonly lake-filled, valleys. The coast is deeply embayed, with narrow peninsulas and numerous islands and shoals.

Above the marine limit, bedrock is not as well exposed. In the northwestern part of the area, good outcrops along lake shorelines and ridge crests are separated by boulder fields. Glacial sediments, including southeast-trending drumlins, obscure bedrock in the north-central part of the area. Several eskers form south to southeast-trending gravel ridges, and one, west of Mistake Creek, has been reworked to form a series of raised beaches.

The area is within the Arctic-Alpine biotic zone. Only in sheltered spots can dwarf birch or willow grow. Away from the coast, outcrops have a heavy lichen cover, with intervening areas covered by moss, grasses, and sedges.

Isolated barren ground caribou were commonly observed on traverse. Arctic hare and arctic fox were seen only rarely. A single polar bear was observed along the eastern shore in 1970, while seven bears were seen inland in late July and August of 1971, apparently crossing the area from west to east. Many of the lakes contain lake trout and arctic char. Ptarmigan are common, while ducks, geese, shorebirds and hawks nest in the area. Seals may be seen on the sea ice before breakup and in the bays during the summer. Whales were seen in Bernheimer Bay in late August 1970.

Acknowledgments

The writer received expert mapping assistance from A.N. LeCheminant in 1970 and 1971 and M.E. Cherry and R.J. Colley in 1971. A. Hickinbotham and R. Buchanan (in 1970) and I. Hutcheon, R. Tremblay, and M. Meunier (in 1971) assisted capably in the field. G.R. Dunning and C. LeTourneau as well as Hutcheon and Cherry aided in office and laboratory projects.

Special thanks are expressed to R. Williamson of the Arctic Research and Training Centre, personnel of the Department of Indian Affairs and Northern Development at Rankin Inlet, and Father Meuss and Brother Paradis of the Roman Catholic Mission at Chesterfield Inlet who provided friendly and invaluable assistance with logistics. C. Bizzard of Aquitane Company, D.W. Halliday of the former Earth Physics Branch, and R. Chewpa of Surveys and Mapping Branch kindly permitted subcharter of their aircraft during fieldwork.

Midwest Airlines provided the helicopter, expertly piloted and maintained by B. McPherson. Thomas Lamb Airways supplied the Cessna 180 with pilot R. Comfort and shared with Transair Limited the job of freighting personnel and equipment. In 1970, Hakaluk kindly agreed to pilot his boat Kakoodluk from Rankin Inlet to Daly Bay, carrying crew and equipment to the field.

Throughout the project, W.W. Heywood unselfishly provided invaluable advice, drawing on his extensive geological experience and expertise in northern logistics.

GENERAL GEOLOGY

The map area lies within the Churchill Structural Province of the Canadian Shield. All bedrock is Precambrian, with the latest metamorphic and igneous events probably occuring during the Aphebian. The oldest rocks (see Map 1652A, in pocket) are gneiss and mafic intrusive rocks of the Daly Bay Complex, which underlies the south-central two-thirds of the map area. In the central part of the complex, the rocks retain granulite facies mineralogy and textures, but towards the margins of the complex they have been strongly sheared, folded, and retrometamorphosed.

Faults and shear zones separate the Daly Bay Complex from surrounding granitic gneiss, amphibolite, and migmatite. Small granite and monzonite bodies intrude the granitic gneiss. Northwest-trending diabase dykes postdate all other lithologies.

Daly Bay Complex (units **A**gl, to **A**gn)

Heywood (1966) defined the Daly Bay Complex as the distinct lithological association of gabbro, anorthosite, generally mafic rich gneiss, and granulites found in the Daly Bay area. The boundary between these rocks and the surrounding gneiss cannot be precisely delineated as it is a complicated metamorphosed shear zone. For purposes of this report, the Daly Bay Complex is thus defined as those rocks which retain either mineralogical or textural evidence of having undergone granulite facies metamorphism. The resulting boundary corresponds closely to Heywood's (1967) map.

Six map units have been chosen to describe the complex. Units **A**gl to **A**gb are the oldest rocks, and these units are homogeneous in lithology and have well defined mutual contacts. Units **A**su and **A**gn represent rocks derived from units **A**gl to **A**gb by a variety of deforma-

[1] Note **A** represents Archean, while A represents Aphebian.

tional and metamorphic processes. These units are lithologically heterogeneous, hence contacts with other units in the Daly Bay Complex are gradational and have been chosen arbitrarily.

Pyroxene granulite (unit Agl)

Pyroxene granulite underlies most of the central portion of the Daly Bay Complex northeast of Bernheimer Bay. These rocks are fine-to medium-grained with diffuse to moderately well developed compositional layers from 1 to 200 cm thick (Fig. 2,3). Well developed 50 to 200 m wide ridges and valleys run parallel to compositional layering. Weathering colour ranges from white to yellowish brown, although extensive lichen growth commonly gives outcrops a dark grey aspect. Most outcrops have a foliation defined by flattening and alignment of mineral grains which is parallel to compositional layering. Locally developed quartzofeldspathic segregations occur as irregular crosscutting patches as well as conformable lenses up to 2 m in length.

Within the area mapped as pyroxene granulite are local layers of quartzofeldspathic granulite **A**qg, and mafic rocks (unit **A**gb). These layers reach widths of 50 m and may extend for hundreds of metres along strike.

Retrograde metamorphic effects are developed sporadically throughout this unit, but are particularly evident along the western and northwestern margins of the large area of pyroxene granulite in the centre of the complex. These effects include growth of secondary chlorite and amphibole as well as local development of microcline porphyroblasts.

This unit includes rocks with a wide variation in abundance of major minerals. Most specimens contain 25 to 50% antiperthitic plagioclase (An_{40} to An_{55}), 5 to 35% quartz, 0 to 15% perthitic microcline, up 20% orthopyroxene, somewhat lesser amounts of clinopyroxene, and up to 10% biotite. As much as 5% hornblende and/or garnet occur locally. Accessory minerals are apatite, zircon,

Figure 2. Weakly layered pyroxene granulite 10 km south of Lake of Islands. Hammer handle 35 cm. (T.M. Gordon, GSC 204068-L)

Figure 3. Layered pyroxene granulite 5 km northeast of head of Bernheimer Bay. Hammer handle 30 cm. (R.J. Colley, GSC 179071)

and opaques. White mica, epidote, and carbonate occur as alteration products of feldspar, and blue-green amphibole, cummingtonite, chlorite, and talc have partially replaced mafic minerals.

Textures range from granoblastic to mylonitic. In a typical rock the pyroxenes are concentrated in individual lamellae, with large ragged crystals being surrounded by smaller grains. Quartz occurs as elongate aggregates of polygonized crystals. Plagioclase forms both a fine grained granoblastic matrix and separate lamellae. Myrmekite, associated with perthitic microcline, is developed sporadically.

The protolith for these rocks is uncertain. The presence of well developed layering and association with quartzofeldspathic granulite **A**qg is consistent with a sedimentary origin. Some areas, however, could be igneous rocks recrystallized in the granulite facies. Large scale isoclinal folding and deformation of all Daly Bay Complex rocks prevents assignment of a stratigraphic position to this unit.

Quartzofeldspathic granulite, paragneiss, marble, and quartzite (units Aqg, Ams)

Much of the northeastern part of the Daly Bay Complex is underlain by garnetiferous quartzofeldspathic granulite and associated metasedimentary rocks (Fig. 4). A weak compositional layering on a scale of 10 to 50 cm is visible in most outcrops. Near Lake of Islands layering is disrupted and the rocks are partially migmatitic and contain zones of granitoid material.

Quartzofeldspathic granulite is white weathering with up to 10% conspicuous 2 to 6 mm reddish garnet porphyroblasts. Paragneiss and schist interlayered with the granulite are commonly graphitic and rusty weathering. Near Lake of Islands textures are granoblastic, but in general rocks have a well developed foliation and a lineation defined by quartz ribbons and lenticles (Fig. 5).

Pyroxene granulite **A**gl and mafic layers **A**gb occur within the area mapped as quartzofeldspathic granulite, but are too small to be mapped.

In thin section, quartzofeldspathic granulite is commonly composed of 20 to 30% quartz, 10 to 40% plagioclase (An_{35} to An_{50}), 5 to 35% perthitic microcline, 0 to 15% poikiloblastic garnet and 0 to 10% biotite. Sillimanite, cordierite, hypersthene and graphite occur in some samples. Accessories are apatite, monazite, zircon, rutile, and spinel. Chlorite and, less commonly, biotite rim and replace garnet porphyroblasts in some localities. White mica, carbonate,epidote, and prehnite are common alteration products. Paragneiss and schist have similar mineralogy, but contain more biotite, sillimanite and garnet.

Minor amounts of marble with associated quartzite, paragneiss and sillimanite schist (unit **A**ms) occur in two restricted areas within the Daly Bay Complex and have been shown separately on the map. Blue-grey weathering impure marble and quartz-rich gneiss occur within layered gneiss along the shores of Bernheimer Bay. The marble occurs as discontinuous 1 to 10 m pods and layers with internal compositional banding produced by light brown calc-silicate-rich zones. The associated metasedimentary rocks are compositionally heterogeneous with 5 to 20 cm layers. Grey, massive garnet-sillimanite schist forms a small island on the west side of Bernheimer Bay and occurs as layers within the adjacent gneiss. At this locality the layering in the schist is contorted but appears to be conformable with surrounding layered gneiss. Small aplite dykes locally intrude the metasedimentary rocks.

Figure 4. Quartzofeldspathic granulite 5 km north of mouth of Borden River. Bell 47 helicopter in background. (T.M. Gordon, GSC 204068-R)

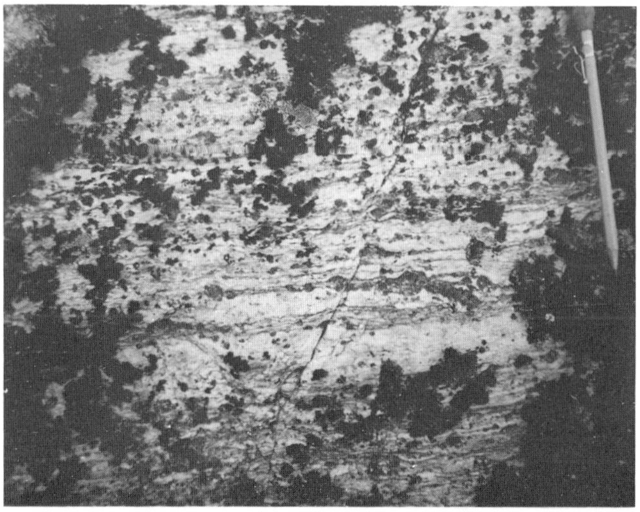

Figure 5. Garnetiferous quartzofeldspathic granulite, showing compositional layering. Black areas lichen. (T.M. Gordon, GSC 204068-P)

Approximately 2 km west of Borden River, at latitude 64°17', an association of marble, orthoquartzite, iron-formation, and paragneiss occurs in a 1 km by 2 km lens-shaped area within unit **A**su. The marble has a blue-grey, lichen-free weathering surface. Orthoquartzite layers up to 10 m thick are boudinaged. These rocks are cut by minor quartz veins and several metamorphosed mafic sills.

Marble commonly has a granoblastic texture and contain diopside and/or tremolite, serpentine after forsterite, and chlorite. Scapolite, potassium feldspar, quartz, and clinozoisite occur in some samples. Quartzite has minor amounts of feldspar, chlorite, white mica and biotite. Quartz grains have strongly sutured boundaries and display undulatory extinction.

The associated paragneiss and schist carry biotite, garnet, sillimanite, and, rarely, cordierite in addition to quartz and feldspar. Secondary minerals are white mica, chlorite, and epidote.

The paragneiss and sillimanite schist of unit **A**ms are indistinguishable from similar rocks interlayered with quartzofeldspathic granulite of unit **A**qg. Although the marble-bearing metasedimentary rocks could represent a separate sedimentary sequence, they probably underwent the same metamorphism and deformation as the quartzofeldspathic granulite. These rocks are thus interpreted to have originated as a single sedimentary sequence including the rocks of unit **A**gl.

Anorthosite, gabbroic anorthosite, layered gabbroic anorthosite (units Aan, Ala)

Several plutons of anorthosite, gabbroic anorthosite, and layered gabbroic anorthosite occur within the Daly Bay Complex. They range from 2 m thick pods and layers to a 30 km long teardrop shaped body east of Daly Bay. Anorthositic rocks commonly form lichen free resistant ridges and rounded outcrops (Fig. 6,7) that are characteristically light grey , while more mafic rocks are less resistant and weather brown. Compositional layers from 15 cm to 2 m thick occur in a few places (Fig. 8). Deformation at the margins of most plutons (Fig. 9) has resulted in gradational and conformable contacts in which slightly sheared rock retaining igneous textures grades into

Figure 6. Rounded resistant weathering hill of anorthosite in large anorthosite pluton east of Daly Bay. (T.M. Gordon, GSC 204068-Q)

Figure 7. Resistant weathering anorthosite dyke at north boundary of Daly Bay Complex. Hammer handle 35 cm. (T.M Gordon, GSC 204068-T)

Figure 8. Layering in gabbroic anorthosite in anorthosite pluton at north boundary of Daly Bay Complex. (A.N. LeCheminant, GSC 179116)

Figure 9. Sheared anorthosite at northeast boundary of large anorthosite pluton east of Daly Bay. (A.N. LeCheminant, GSC 179058)

strongly foliated plagioclase-hornblende gneiss indistinguishable from country rock.

Anorthosite dykes, 1 to 10 m wide and up to 10 km long occur in the Boundary Shear Zone (Asu). They are monomineralic and generally display mylonitic textures.

Gabbro and gabbroic anorthosite plutons lying north of Lake of Islands and near the mouth of Borden River locally crosscut layering in the enclosing gneiss. Compositional layering is common and these bodies retain much of their primary mineralogy. The rocks are medium grained with allotriomorphic granular textures somewhat modified by deformation. Plagioclase grains, An_{50} to An_{80}, have sutured and polygonized boundaries and commonly have curvilinear and wedge-shaped twins. Orthopyroxene, clinopyroxene, and hornblende are the dominant mafic minerals. Accessories are red-brown biotite, opaques, and apatite. Secondary minerals are serpentine and cummingtonite after orthopyroxene, as well as green biotite, chlorite, sericite, and prehnite.

The large anorthosite plutons at the western and northern boundaries of the complex have a crude internal zoning. The central portions are coarse grained, massive, and little deformed with relict igneous textures, in which feldspar laths up to 10 cm in length enclose intercumulate mafic patches pseudomorphic after pyroxene. Towards the margins, layering becomes more prominent (unit **A**la), although it is partially obscured by an increase in deformation which is reflected in development of a strong marginal foliation.

Plagioclase compositions in these anorthosites range from An_{40} to An_{55}. In less deformed rocks the mafic minerals display corona structures. From core to rim, the sequence of such structures is: orthopyroxene, or rarely, clinopyroxene; cummingtonite; blue-green hornblende — quartz symplectites; garnet and/or biotite; plagioclase. Orthopyroxene and cummingtonite are commonly dusted with opaques. Not all corona structures contain all zones. Pyroxenes are commonly completely replaced. Cummingtonite- hornblende and hornblende-quartz symplectites are the most common mafic aggregations observed.

Minor amounts of prehnite, biotite, and sericite occur within plagioclase grains in most specimens. Opaque minerals occur sporadically, and are sometimes rimmed by garnet. Some plagioclase is saussuritized and chlorite, epidote, and carbonate embay and replace amphiboles.

Deformation and metamorphism have obscured most contact relationships of these rocks with surrounding gneiss. Local crosscutting relationships together with the igneous textures and compositional layering observed in some bodies, indicate that the anorthositic rocks were intruded as magmas into the rocks of units 1 and 2. The mylonitic anorthosite dykes probably represent the remains of tectonically disrupted bodies.

Gabbro, metagabbro, and amphibolite (unit **A**gb)

Subconcordant dykes and sills of gabbro, metagabbro and amphibolite occur throughout the Daly Bay Complex. They range from 2 to 100 m in width and are up to 10 km long (Fig. 10). Only the large plutons are shown on the map. Mineralogical layering was observed in some of these plutons, but generally the rocks have a uniform texture. Although these intrusions have been tectonically disrupted and deformed, they often retain sharp crosscutting boundaries with country rock and do not have the intense marginal shearing characteristic of the anorthosite suite (unit **A**an).

Plagioclase, hornblende, and clinopyroxene are the most common minerals. Red-brown biotite and opaques as well as minor amounts of quartz and microcline are present in most amphibolites between Bernheimer Bay and Borden River. Garnet is common in the amphibolites and metagabbros at the northern and western edges of the complex. Although some igneous textures are retained, most samples have a granoblastic texture. Secondary chlorite, epidote, and sericite occur in all thin sections examined.

Figure 10. Deformed mafic layers in Boundary Shear Zone on coast 10 km east of large anorthosite pluton at west margin of Daly Bay Complex. 30 cm hammer at right centre of photograph. (T.M Gordon, GSC 204068-O)

These mafic bodies probably were intruded as dykes and sills after the beginning of deformation of the complex but prior to cessation of movement and amphibolite facies metamorphism.

Boundary Shear Zone (units **A**su, **A**sf, **A**sm)

A heterogeneous assemblage of highly deformed, strongly foliated layered gneiss, augen gneiss, and mylonite forms a 1 to 15 km wide Boundary Shear Zone to the Daly Bay Complex. Most of these rocks are mineralogically equivalent to less deformed rocks in the central part of the complex, but are intermixed in lenses and layers too small to be mapped separately. Areas in which a particular lithology is dominant are indicated by individual symbols (**A**sf, quartzofeldspathic rocks; **A**sm, mafic rocks) on the accompanying map.

Many outcrops have a distinctive banded appearance resulting from a 2 mm to 2 m compositional layering that is consistent in width and orientation for hundreds of metres along strike (Fig. 11). At several localities in the western part of the complex, layered gneiss has been deformed to produce complex fold patterns (Fig. 12). In other outcrops, boudinage and shears disrupt layering (Fig. 13, 14). Across strike textures show rapid transitions between mylonite, augen gneiss, and fine grained granoblastic gneiss. Local retrograde metamorphism of mafic rocks gives a sooty appearance to mafic minerals and a bluish cast to quartz. In felsic rocks a strong lineation is defined by quartz ribbons 2 to 5 mm wide and up to 20 cm long.

Aplite veins and stringers, and irregular pegmatite pods up to 2 m in maximum dimension occur throughout this unit. They show some evidence of internal deformation but crosscut the strong mylonitic foliation. Mafic rocks of unit **A**gb form concordant but boudinaged sills within unit **A**su.

Figure 11. Layering in Boundary Shear Zone. Note consistency of foliation attitude and layer thickness. Hammer in foreground 30 cm, assistant in background 170 cm. (T.M. Gordon, GSC 204068-K)

Figure 12. Thin layered mylonite in Boundary Shear Zone (unit 5) immediately west of large anorthosite at west boundary of Daly Bay Complex. Hammer 30 cm. (T.M. Gordon, GSC 191792)

In thin section, these rocks are seen to contain similar minerals to rocks in the core of the complex, although in places deformation and local retrometamorphic overprinting have partially to completely altered rock textures and mineralogy. Clinopyroxene and orthopyroxene occur as porphyroclasts or as ragged, polycrystalline lamellae and lenses. Alteration of pyroxene to chlorite and various amphiboles occurs locally. Garnet shows partial to complete replacement by biotite and/or chlorite. Potassium feldspar occurs as augen and porphyroclasts with polygonized boundaries. Alteration products of feldspar are sericite and prehnite.

Figure 13. Augen and boudinage in Boundary Shear Zone 10 km northwest of Whale Point. (R.J Colley, GSC 179101)

Figure 14. Laminated mylonite in Boundary Shear Zone on coast 5 km south of the mouth of Mistake Creek. (T.M Gordon, GSC 204068-V)

This unit has gradational contacts with rocks inside the complex. For mapping purposes this has been taken to be the locus of major faults, even though the unit itself contains many faults and mylonite zones. The outer contact of the boundary shear zone is relatively abrupt. Along the western and northern edge of the complex is a curvilinear fault zone, approximately 100 m wide, containing both mylonite and chloritic phyllonite. Although rocks north and west of this fault are commonly thinly layered, in part cataclastic, and have a foliation parallel to the fault, mafic lithologies apparently contain no pyroxene, and garnet is rare. Between Whale Point and Borden River the contact is not well exposed, but is assumed, on the basis of published aeromagnetic data (Geological Survey of Canada, 1973) to be equally abrupt.

This unit is interpreted to be a ductile shear zone that has involved rocks of units **A**gl to **A**an, and has been intruded by rocks of unit **A**gb late in the deformation history. Retrograde metamorphism is more consistently developed in the Boundary Shear Zone than in the core of the Daly Bay Complex, but is does not have a sufficiently regular distribution to permit mapping of a retrograd.

Layered gneiss and migmatite (units **A**gn, **A**mg)

Garnet, hornblende and biotite bearing layered gneiss (unit **A**gn) and migmatite (unit **A**mg) underlie a 10 km wide belt centred on Bernheimer Bay. These layered rocks are similar to the gneisses of unit **A**su, but commonly have more contortion and disruption of layering. Many outcrops contain augen, quartz ribbons, and other evidence of ductile deformation, but these features are themselves recrystallized and deformed (Fig. 15). Similar rocks occur in isolated areas within the Boundary Shear Zone, particularly in the northwest part of the Daly Bay Complex.

Granodiorite to tonalite veins, dykes, and sills cut these layered rocks. With increasing amounts of granitoid material this rock grades into zones of migmatite and agmatite (unit **A**mg).

The layered gneiss is compositionally and mineralogically heterogeneous, ranging from plagioclase — hornblende — quartz — biotite mafic gneiss to quartz — microcline — plagioclase — garnet — sillimanite gneiss. Retrograde metamorphism is ubiquitous. Fine grained amphiboles, biotite, and chlorite commonly rim or replace garnet and pyroxene, giving most of the mafic rocks a dull, sooty appearance. In many localities, quartz has a bluish cast.

Retrograde effects, visible in outcrop, are reflected in microscopic textures and mineralogy. In addition to the features noted above, secondary chlorite and white mica were observed in several samples. Secondary biotite and euhedral epidote are particularly evident in rocks from the north end of Bernheimer Bay.

Contacts of this unit with adjoining rocks are gradational and have been arbitrarily drawn to coincide with a change in topographic character of the terrain. The presence of relict deformational textures and high grade mineral assemblages suggest that these rocks were derived from rocks of units **A**gl to **A**su by rehydration and retrometamorphism under amphibolite facies conditions.

Other map units

Layered biotite gneiss, hornblende gneiss, amphibolite, and migmatite (units Agn to Amg)

The rocks surrounding the Daly Bay Complex are mainly various layered gneisses and migmatite. Although they are intermixed and grade into each other, an attempt has been made on the map to identify areas where a particular lithology is dominant.

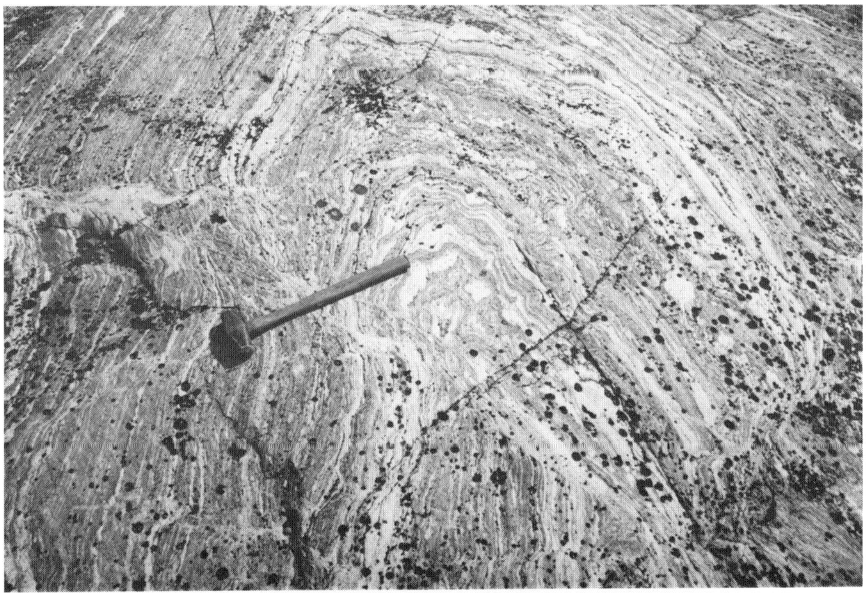

Figure 15. Deformed biotite-rich augen gneiss in map unit 6 at Cape Fullerton. (T.M. Gordon, GSC 204412-B)

Biotite-bearing layered gneiss (unit Abg) weathers in various shades of pink and grey and has a weak to moderately well developed mineralogical layering, commonly with a parallel foliation. Layers range in thickness from one centimetre to several metres and are usually contorted and disrupted (Fig. 16).

These rocks are medium grained, granoblastic, and mineralogically simple. Oligoclase is the dominant feldspar, with potassium feldspar forming up to 30% of some rocks. Quartz content ranges from 5 to 25%. Various proportions of green biotite and hornblende with subsidiary epidote comprise the mafic minerals which may attain amounts of 20% in some samples. Garnet occurs in some amphibolites. Sphene, apatite, and opaques are ubiquitous accessories. Secondary alteration is particularly common near northeast-trending fault zones, resulting in the development of chorite, sericite, and carbonate.

Hornblende gneiss (unit Ahg) and amphibolite (unit Aam) are particularly common in the southwestern and northeastern parts of the map area (Fig. 17). Amphibolite layers up to 500 m thick are commonly foliated and include both massive and compositionally layered types. Several large amphibolite bodies occur within the Outer Shear Zone (unit As) along the north and northeast flanks of the Daly Bay Complex. Most amphibolite probably originated as mafic igneous rock, emplaced prior to the regional amphibolite facies metamorphism.

Areas mapped as migmatite (unit Amg) include both rocks with diffuse and highly disrupted layering as well

Figure 16. Layered biotite gneiss on island 6 km west of southern tip of large anorthosite pluton at western margin of the Daly Bay Complex. (T.M. Gordon, GSC 204068-Z)

Figure 17. Layered hornblende gneiss at southwest margin of map area. Field of view is approximately 200 m. (T.M. Gordon, GSC 204068-W)

as agmatitic varieties in which angular blocks of layered gneiss are enclosed in massive granodiorite to tonalite.

Rocks of this unit are in fault contact with the Daly Bay Complex and grade into the thin layered gneiss of unit As. The possibility that they are in part retrometamorphosed equivalents of Daly Bay rocks cannot, however, be entirely discounted. Local similarities to the layered gneiss and migmatite at Bernheimer Bay suggest that deformed and completely recrystallized Daly Bay Complex rocks would be indistinguishable from many of the rocks of unit Agn.

Outer Shear Zone (units As, Asc, Asq)

Several zones of thin layered gneiss, schist, and mylonite occur within the layered gneisses of unit Agn and adjacent to the Daly Bay Complex. The largest of these forms a 1-5 km wide belt roughly concentric with the complex, while a smaller belt trends northeasterly from the head of Daly Bay. Similar rocks occur scattered throughout the northern part of the map area.

Compositional layers within gneissic rocks of this unit are chiefly only 2 to 20 mm thick. A well developed foliation due to parallel alignment of platey minerals is common. These features are best developed in the northern part of the shear zone where they impart a characteristic strong fissility to the rocks and result in more recessive weathering outcrops than in other map units. At the western and eastern margins of the Daly Bay Complex the rocks are not as well foliated, although they retain their thin layered character.

Mineralogically, the gneisses are identical to the gneisses of unit Agn. Oligoclase, quartz, and orthoclase occur in proportions ranging from monzonitic through granodioritic to tonalitic. Mafic minerals range from trace amounts to 10% and consist of various proportions of green biotite, hornblende, chlorite, and epidote. White mica, apatite, sphene, and opaques are accessories.

Thin zones of mylonite, augen gneiss, and cataclasite (unit Asc) occur at several places within the shear zone. These rocks are dark grey, pink, and light green weathering and laminated, but generally do not have a pronounced compositional layering. In thin section, cataclasite has a microbreccia fabric and well developed low grade alteration in which clay minerals replace feldspars and mafics.

A 10 km long belt of fine grained thin layered gneisses trends northeast from the head of Daly Bay. The foliations in this belt define a synform which is cored by a complex assemblage of impure quartzite, marble, and ultramafic rock (unit Asq). Isolated 1 to 20 m blocks of marble and ultramafic rock are also present in gneisses along strike from this belt. The association of marble and quartz ite suggests a lithological correlation with similar rocks in unit Ams of the Daly Bay Complex. This synform may represent an allochthonous tectonic sliver within the Outer Shear Zone.

These shear zones grade abruptly into the foliated gneisses of unit Agn and the mylonitic rocks of the Boundary Shear Zone of the Daly Bay Complex. This zone probably includes some retrometamorphosed granulite facies rocks as well as rocks from the surrounding amphibolite facies gneisses.

Granite and granodiorite (unit Agr)

Veins, sills, and dykes of light grey to pink weathering granite and granodiorite are particularly common in the layered gneiss at the eastern edge of the map area. Most are less than 10 m in width but one pluton, immediately adjacent to the northwest boundary of the Daly Bay Complex, is large enough to be mapped.

This pluton forms an resistant weathering ridge of foliated, pink weathering biotite granite. Elongate xenoliths of fine grained hornblende gneiss are common throughout. Mineralogically, it is a biotite-oligoclase granite. Accessories are sphene, apatite, opaques, and zircon, as well as an unidentified metamict mineral. The feldspars show strain twinning, while quartz has undulose extinction and sutured boundaries. Some fine grained quartz aggregates produce local mortar textures.

The granite is concordant with the gneiss at its northwestern border, but aplite veins in the vicinity crosscut the enclosing rocks. At the southern boundary of the pluton the granite intrudes rocks of the Boundary Shear Zone of the Daly Bay Complex.

Pyroxene monzonite (unit Amz)

Two plutons of coarse grained pyroxene monzonite intrude layered gneisses in the northern part of the map area. They form roughly equant stocks 1-2 km in diameter. Contact relations of the eastern stock are unknown but the western one, which has a massive core with a foliated margin, is conformable with the surrounding gneiss.

A sample from the eastern stock has a composition of 10% quartz, 50% plagioclase (An_{30}), 30% perthite, and 10% mafics. The mafic minerals occur as ragged fine grained aggregates interstitial to the coarser grained quartz and feldspars. Orthopyroxene grains with prominent exsolution blebs of clinopyroxene are intergrown with irregular grains of hornblende and oxides. Garnet occurs as polycrystalline rims, partially to completely surrounding the intergrowths, although garnet is usually separated from orthopyroxene by a thin quartz selvage. Randomly oriented blades and sprays of biotite intergrown with quartz appear to be associated with garnet. A sample from the margin of the western stock contains clinopyroxene rather than orthopyroxene. Textures in this sample are dominated, however, by a overprinted tectonic fabric.

Textures within these rocks suggest that pyroxene was a liquidus phase and that retrograde metamorphism to produce garnet, hornblende, and biotite took place after crystallization. The stocks appear to have been in part forcibly emplaced during the waning stages of the regional amphibolite facies metamorphism.

Diabase (unit Hdb)

Several northwest trending diabase dykes cross the map area. The dykes form brown weathering resistant ridges 2-50 m in width. These rocks are composed of zoned labradorite laths with intersertal augite grains. Quartz-feldspar myrmekite and opaques occur in minor amounts. The feldspar is locally saussuritized, whereas the pyroxenes show slight alteration to actinolite.

These dykes represent the youngest igneous event recognized in the area. They are part of the Mackenzie swarm which has been assigned an age of 1200 Ma by Fahrig and Jones(1969).

STRUCTURAL GEOLOGY

Structures within the map area are dominated by features reflecting the emplacement of the granulite facies rocks of the Daly Bay Complex into the surrounding amphibolite facies gneiss. Mesoscopic structures and fabrics show a regional zonation inferred to have resulted from a zonation in deformational regimes. Within the core of the Daly Bay Complex, ductile deformation is suggested by continuity of layering and large scale folds. In places, however, mineral textures are granoblastic, suggesting annealing at granulite grade. At the margins of the Complex, conditions were such that planar fabrics and finer grain sizes are preserved.

Outward from the Daly Bay Complex, deformation under amphibolite facies conditions produced bands of thin layered and strongly foliated gneiss alternating with less well foliated rocks. Later brittle deformation produced local cataclastic zones with greenschist facies mineralogy. The interaction of metamorphism and deformation means that some rocks probably passed through all metamorphic and deformational regimes, but retain evidence of only the final stages of their journey.

Daly Bay Complex

On a regional scale, structures within the complex define an upright, shallowly southeast-plunging synform with a steep southwest limb and a moderately dipping northeast limb. This trend is at a high angle to the northeast regional trends of this part of the Churchill Province. Major shear zones outline, or are parallel to, the margins of the Daly Bay Complex, indicating that it is allochthonous. All rocks in the Daly Bay Complex have undergone ductile deformation. The amount of strain, as reflected in grain size, layer thickness, and development of mylonite textures, increases from the core of the complex to its boundaries.

Compositional layering is the oldest mesoscopic planar fabric within the complex. Most of it is probably transposed sedimentary bedding or igneous layering, but at some localities it may have resulted from tectonic attenuation of sills, dykes, or metamorphic segregations. Layers vary from lamellae one centimetre in thickness to ridges tens of metres in width. Layering is particularly well developed between Bernheimer Bay and Whale Point. With the exception of major fold closures discussed below, attitudes of layering are persistent along strike.

Superimposed on compositional layering is a foliation defined by parallel alignment of platey minerals and mineral aggregates. This foliation is developed to varying degrees and is generally parallel to compositional layering. Near major fold noses, however, it transects layering, indicating that it developed with an axial planar orientation. A mineral lineation commonly accompanies the foliation. Both are well developed in quartzofeldspathic granulite which commonly contains quartz ribbons.

Mesoscopic folds of compositional layering occur at a few localities. In general they are tight to isoclinal symmetrical folds with steeply dipping axial planes and shallow plunges. Changes in attitudes of layering across strike suggest that such folds are common but shallow plunges and flat topography reduce the opportunities for observing such structures in cross-section.

The geometry of mesoscopic structures is summarized in Figure 18. Layering and foliation within the core of the complex dip moderately to steeply west. Lineations plunge to the south in the northern part of the area but are generally shallow to north plunging in the south. Exceptions to the pattern occur mainly near the mouth of the Borden River where south plunging mineral lineations dominate.

Two, large scale, doubly plunging tight to isoclinal synforms deform large scale layering of pyroxene granulite between Bernheimer Bay and Lake of Islands. The axial traces of these folds trend northwest and extend for approximately 20 km. The easterly of the two is an upright fold. Metamorphic overprinting and later deformation partially obscure the western synform. It is overturned to the east throughout its length, with a moderate to steeply dipping eastern limb.

There are insufficient data to determine whether the doubly plunging nature of these folds is a result of several deformational episodes or whether they initially formed with their present geometry. The absence of antiformal structures of comparable size is a weak argument against the former hypothesis.

Three kilometres north of the mouth of Borden River large scale compositional layering in quartzofeldspathic granulite outlines a major fold closure. Mineral lineations in the vicinity plunge shallowly northwest, suggesting that the fold is a synform overturned to the northeast. Its axial trace extends from the coast northwest for 30 km where it is truncated by a fault.

These early structures cannot be traced beyond the southeast part of the complex. On the west, at Bernheimer

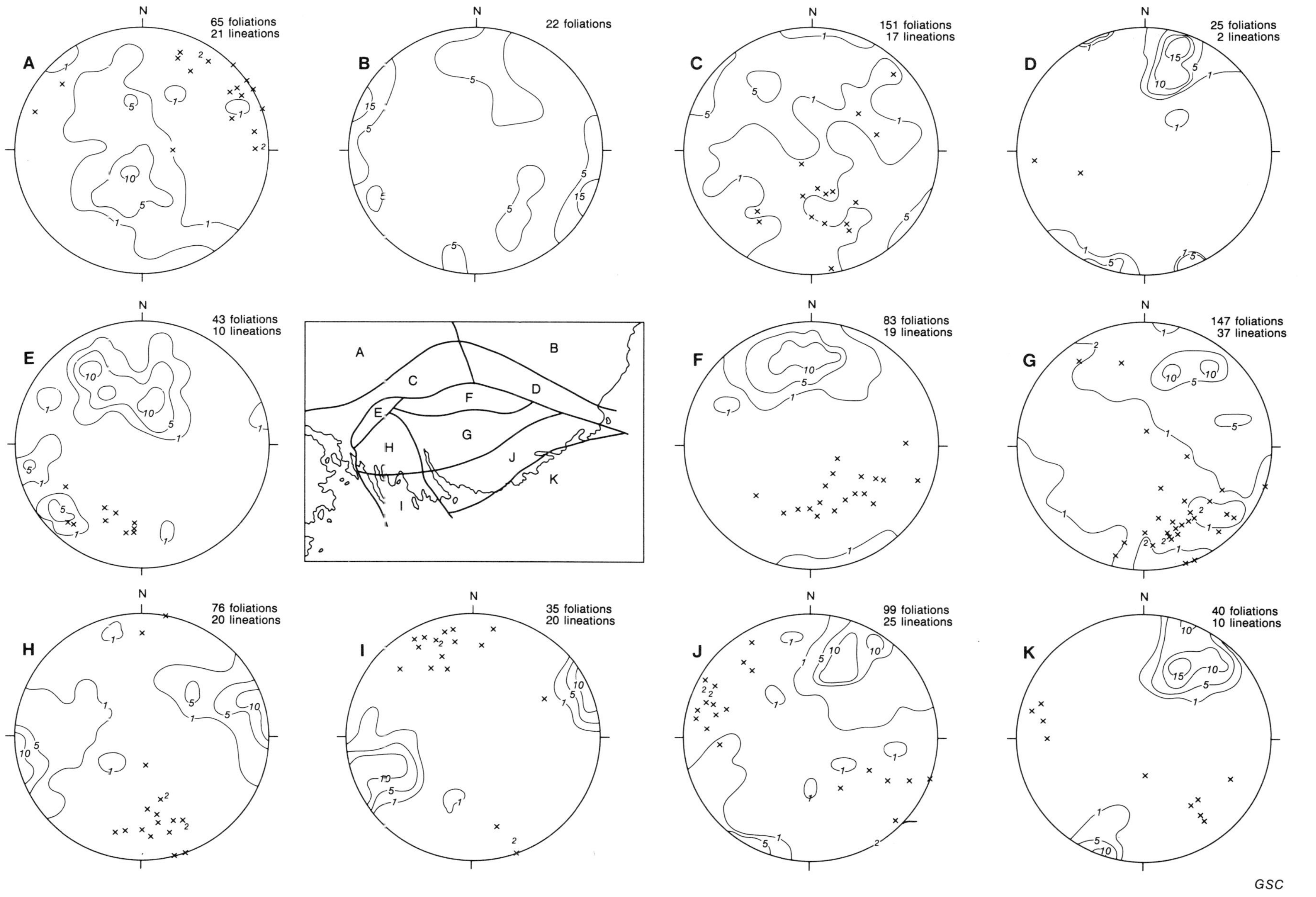

Figure 18. Geɔmetry of mesoscopic structures. Fabric elements plotted on lower hemisphere of an equal area projection. Gneissosity and foliation poles are contoured, contours labelled in percent per one percent area. Mineral lineations shown by crosses.

Bay, amphibolite facies metamorphism and migmatization disrupt and obliterate the earlier trends. A fault truncates the eastern limb of the easternmost fold at a small angle, although attitudes within the Boundary Shear Zone east of the fault are subparallel to those within the core of the complex. Towards the north, structural trends swing from northwest to west. Trend lines, determined from aerial photographs, suggest a number of fold closures in this region but their vergence cannot be resolved from the available ground observations.

The oldest faults observed within the Daly Bay Complex are those associated with the inner margin of the Boundary Shear Zone. They are recognized by abrupt across-strike changes in abundance of mylonite and augen gneiss or by truncation of layering and foliation. The fault surfaces are presumably subparallel to foliation in adjacent rocks with motion parallel to locally developed lineations. No data were obtained that might indicate the sense or amount of displacement. No obvious change in mineralogy is associated with these faults, hence they postdate the initial development of ductile fabrics at granulite facies conditions, but predate retrograde amphibolite facies metamorphism.

A younger set of northeast-trending faults offset layering in gneisses both east and west of Bernheimer Bay. Associated minor faults are vertical, with vertically plunging drag folds indicating sinistral strike slip displacement. The apparent offset is small. Chlorite, epidote, and white mica are developed in schists related to some of these faults, while others contain undeformed pegmatite bodies.

Other late faults , trending north-northeast to north-northwest, were identified on air photographs. No dips or sense of motion have been determined for these. They are presumed to be subvertical with minor displacements.

Boundary Shear Zone

Fabrics in the Boundary Shear Zone are similar to those in the core of the Daly Bay Complex but preserve more evidence of ductile strain. Layering and minor folds are much more strongly attenuated, while augen gneiss, mylonite, and boudinaged layers are abundant. Foliation is everywhere parallel to compositional layering. A strong quartz ribbon lineation is characteristic of quartzofeldspathic rocks.

Mineral lineations developed along the northern margin plunge moderately to steeply southwest to southeast. Along the east and west margins, similar lineations have dominantly shallow south plunges in the north and dominantly shallow north plunges in the south. This extension direction suggests that the Daly Bay Complex was emplaced parallel to the present northeast-trend.

In a several places foliation/layering surfaces are folded. Both symmetric isoclinal folds with axes parallel to the local lineation and complex irregular folds with oblique axes are present. Minor folds have no consistent sense of asymmetry.

The north and west boundary of the Boundary Shear Zone is a curvilinear fault zone dipping steeply to the south and east and containing mylonite and chloritic phyllonite. This zone is parallel to foliations inside and outside the complex. The fault was presumably active during the granulite facies metamorphism affecting Boundary Shear Zone rocks as well as during later greenschist facies conditions. Movement on the fault was probably parallel to adjacent south plunging mineral lineations. No definitive criteria for sense of movement were obtained but it is most likely that the higher grade rocks of the Daly Bay Complex to the south were emplaced upwards into the surrounding amphibolite facies gneiss to the north.

Outer Shear Zone

The Outer Shear Zone forms a 5 to 10 km wide belt surrounding the Daly Bay Complex. Rocks within the shear zone have thin compositional layering parallel to a well developed foliation, which is defined by alignment of biotite and tabular mineral grains. Attitudes are generally conformable with trends in the adjacent granulite facies rocks of the Boundary Shear Zone.

Microscopic textures range from protoclastic to mylonitic. Mortar textures are most common in the gneisses. Angular to subrounded feldspar and quartz grains have undulatory extinction and polygonized boundaries. Chlorite and white mica occur both as subrandomly oriented discrete grains and as anastomosing surfaces which both rim and transect quartz and feldspar grains. Cataclasite (unit As) has a microbreccia fabric.

At the north margin of the Daly Bay Complex, between the Outer Shear Zone and the Boundary Shear Zone, is a sigmoidal wedge of hornblende gneiss and amphibolite (unit Agn). These rocks are not as intensely strained as rocks in the adjacent shear zones. Layering and foliation are sufficiently well developed, however, to indicate a southwest plunging synformal structure with an overturned southern limb. This structure is interpreted as a preserved infold of country rock. The apparent asymmetry of the fold is consistent with reverse movement on the fault forming the northern boundary of the Daly Bay Complex.

Surrounding area

At distances greater than 20 km from the Daly Bay Complex, the main mesoscopic fabric element is compositional layering, chiefly defined by variable concentration of mafic minerals. Although some layering may be due to original sedimentary or igneous layering, it is highly probable that many of the rocks have recrystallized from highly strained precursors and that the layering is due to tectonic attenuation of pre-existing compositional inhomogeneities with a variety of origins.

Foliation is not as well developed as in the Outer Shear Zone, but where present, it is parallel to compositional layering. Mesoscopic folds are highly variable in style and orientation, although the overall structural trend is northeast. Migmatitic and agmatitic zones contain disrupted layering and foliation, with no discernable pattern.

Air photo examination reveals a particularly well developed type 2 (Ramsay, 1967) fold interference pattern in the northeast part of the map area. The pattern is outlined by amphibolite layers in hornblende gneiss. Elsewhere, lithological contrast is insufficient to reveal such structures.

In the northwest corner of the map area, layering, foliation, and trend lines suggest a shallowly northwest plunging, generally antiformal structure. The southeastern limb of this structure is truncated by a fault or faults extending from the head of Daly Bay to the north-central part of the map area. South of the fault, attitudes of layering and foliation have moderate to steep northwest dips (in the southwest) and moderate southeast dips (in the north). The displacement on this fault is unknown, although lineations in the vicinity plunge shallowly northeast. A chloritic phyllonite has been developed along the fault trace.

In the extreme northeast corner of the map area, a north- northwest-trending escarpment with up to 50 m of relief is interpreted as a near vertical fault with northeast side down displacement. It is probably part of an extensive post-middle Silurian fault system affecting rocks 75 km to the east on Southampton Island (Heywood and Sanford, 1976). Heywood (1967) mapped several faults with the same trend over much of the region to the north. Faults with a similar trend displace layering and foliation in the Outer Shear Zone.

Geophysical constraints

Gibb and Halliday (1975) computed a crustal model for the Daly Bay Complex based on gravity data and densities obtained from rocks collected during the gravity survey. Their model, which includes local compensation, suggests a regional crustal thickness of about 37 km, with a slightly thicker root under the complex. This contrasts with the 23 to 27 km estimates of Mohorovic depth for Baker Lake and Chesterfield Inlet (Ruffman, 1969).

METAMORPHISM

Daly Bay Complex

Rocks in the Daly Bay Complex have been metamorphosed in the granulite facies. Within the complex, there is no recognizable gradient in metamorphic conditions. At the margins of the complex, and near Bernheimer Bay, minerals such as garnet and pyroxene have been partially or completely replaced by hydrous minerals such as biotite, chlorite, and cummingtonite. Near late faults, chorite — epidote — white mica assemblages indicate greenschist facies conditions.

Estimates of maximum grade attained by Daly Bay rocks may be obtained by comparing observed mineral assemblages with various petrogenetic grids and from thermodynamic analysis of compositions of coexisting minerals. Such estimates require the assumption that the assemblages and mineral compositions reflect equilibrium at peak metamorphic conditions. The ductile deformation and variable retrograde metamorphism experienced by these rocks make interpretation of textures difficult, but the assemblages cited below probably have crystallized at equilibrium. The estimated pressure and temperature are interpreted as representing the environment of Daly Bay Complex rocks prior to their tectonic emplacement into the surrounding terrane.

In aluminous rocks the maximum mineral assemblage observed in a single thin section is: quartz — microcline — oligoclase — biotite — garnet — cordierite — sillimanite. Spinel, graphite, apatite, and oxides are accessories in this assemblage as well as in sub-assemblages of the same major minerals. Not all mineral pairs are in contact, but examination of several thin sections suggests that these minerals coexisted in equilibrium. The critical assemblages orthopyroxene — sillimanite — K feldspar — quartz and orthopyroxene — cordierite — K feldspar — quartz are not present. The only aluminosilicate polymorph that has been detected is sillimanite.

These observations suggest that the rocks were metamorphosed at temperatures and water pressures below the quartz — plagioclase — K feldspar — aluminosilicate — biotite — garnet — cordierite — orthopyroxene invariant point of Grant(1985); and at temperatures exceeding the stability of biotite and aluminosilicate in the presence of quartz, plagioclase, ilmenite, and magnetite. The position of the invariant point is estimated at 775°C and 5 kb (0.56 GPa) by Grant (1985).

In mafic rocks, the maximum phase assemblages are: quartz — plagioclase — orthopyroxene — clinopyroxene — biotite — hornblende — K feldspar — opaques — apatite; and quartz — plagioclase — orthopyroxene — clinopyroxene — biotite — garnet — K feldspar — opaques — apatite. In the latter assemblage garnet and clinopyroxene are not in contact. Orthopyroxene — K feldspar assemblages occur throughout the Daly Bay Complex, whereas hornblende — garnet — biotite and clinopyroxene — garnet — biotite assemblages are restricted to the Boundary Shear Zone. Garnets in clinopyroxene bearing rocks commonly have coronas of intergrown feldspar and orthopyroxene.

These observations suggest that Daly Bay Complex rocks were metamorphosed at temperatures and pressures above the quartz — plagioclase — K feldspar — biotite — garnet — hornblende — orthopyroxene — clinopyroxene invariant point of Froese and Jen (1979). Abbot (1982) placed his analogous [Fay,Epi] invariant point at a temperature near 750°C and pressures between 3.5 and 14 kb (1.4 Gpa).

Figure 19 (from E. Froese *in* Schau et al., 1986) illustrated schematically the P-T field implied by the mineral assemblages.

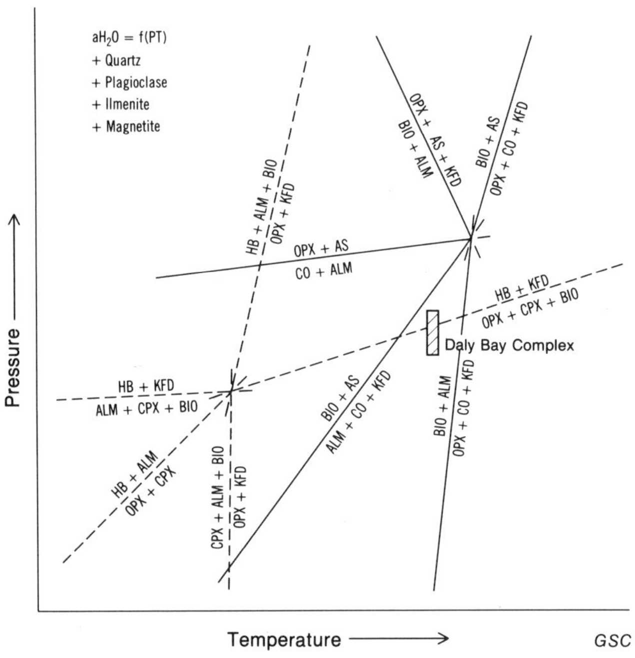

Figure 19. Qualitative petrogenetic grid (E. Froese as reported in Schau et al., 1986). Conditions in the Daly Bay Complex in shaded area.
Abbreviations: ALM — almandine, AS — aluminosilicate, BIO — biotite, CO — cordierite, CPX — clinopyroxene, HB — hornblende, KFD — potassium feldspar, OPX — orthopyroxene.

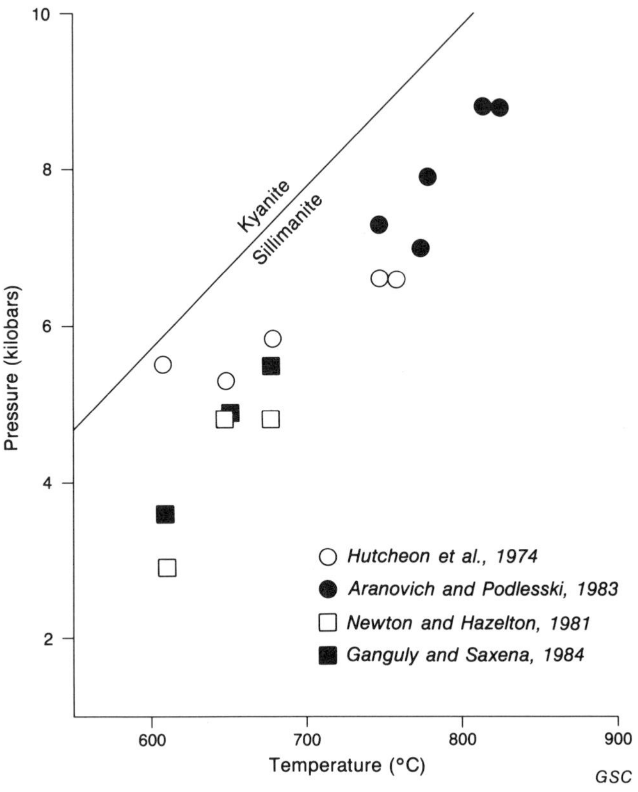

Figure 20. Pressure-temperature conditions in the Daly Bay Complex as estimated by various authors. Kyanite-sillimanite boundary from Holdaway (1971).

Several workers have obtained estimates of pressure and temperature based on compositions of coexisting minerals in samples from Daly Bay (Fig. 20). Hutcheon et al. (1974) used the assemblage quartz — sillimanite — garnet — cordierite and empirically estimated thermochemical parameters to obtain temperature values ranging from 610 to 750°C and pressures between 5.3 and 6.6 kb (0.53-0.66 Gpa). The same mineral compositions were used by Aranovich and Podlesski (1983) to obtain temperature estimates ranging from 750 to 825°C and pressure estimates ranging from 7 to 8.8 kb (0.7-0.88 GPa). Application of the garnet — plagioclase — Al_2SiO_5— quartz barometer to the same samples by Newton and Haselton (1981), gave pressure estimates of 2.9 to 4.8 kb (0.29-0.48 GPa). Ganguly and Saxena (1984) revised Newton and Haselton's (1981) work and obtained pressures between 3.5 and 5.5 kb (0.35-0.55 Gpa). Both Ganguly and Saxena (1984) and Newton and Haselton (1981) accepted the temperature estimates of Hutcheon et al (1974). The higher temperatures obtained by Aranovich and Podlesski (1983) would increase these authours' pressure estimates to between 6 and 7 kb (0.6-0.7 GPa).

Present data thus indicate that rocks of the Daly Bay Complex were metamorphosed at 750 ± 50°C and 0.55 ± 0.1 GPa. They reached these conditions prior to and during the initial stages of their emplacement into the adjacent rocks.

Surrounding rocks

Assemblages in rocks surrounding the Daly Bay Complex are various subassemblages of quartz — plagioclase — K feldspar — biotite — hornblende — epidote, or quartz — plagioclase — hornblende garnet. Such rocks are stable over a wide range of metamorphic conditions in the upper greenschist and amphibolite facies. No gradient in metamorphic conditions was detected in the area. Local chlorite — epidote — white mica assemblages are associated with the latest faulting.

Timing of metamorphism

Rocks within the Daly Bay Complex experienced granulite facies metamorphism both prior to and during ductile deformation. Evidence for this comes primarily from microscopic textures in which orthopyroxene and garnet porphyroblasts show the effects of marginal granulation and disruption, but have not been recrystallized to lower grade minerals. As the deforming rocks passed into a lower grade metamorphic regime, granulite facies minerals were overprinted with hydrous minerals such as biotite, hornblende and cummingtonite. These later minerals have a preferred orientation in some localities, but in others are random, suggesting that during amphibolite facies metamorphism deformation was restricted to discrete zones. Greenschist facies conditions were reached as the brittle regime was encountered, so that assemblages associated with late faults and cataclasite include chlorite and white mica.

GEOCHRONOLOGY

A limited amount of geochronological data is available for the Daly Bay Complex. A uranium-lead analysis of zircon from a two pyroxene enderbitic granulite south of Lake of Islands gave a concordant age of 2067 Ma. The zircons are prismatic but have rounded interfacial angles, consistent with partial resorption under high grade metamorphic conditions. More detailed work has revealed an inherited component of probable Archean age within the zircon under study. This date can thus be interpreted as a minimum age for protolith to the gneisses, and a maximum age for granulite facies metamorphism. A potassium-argon date from biotite in a pegmatite cutting the foliation in the Boundary Shear Zone is 1951 ± 26 Ma, providing a minimum age for cessation of ductile deformation at the margins of the Daly Bay Complex.

Park (1973) studied paleomagnetism of rocks from the Daly Bay Complex and Melville Peninsula areas and reported that a pole from intermediate and basic gneisses lies on the apparent polar wander curve for the Canadian Shield in the time range 1425 to 1725 Ma. This range is comparable to many K-Ar age determinations in northeastern District of Keewatin and presumably reflects the time at which the rocks cooled below 600°C.

Schau (1980) and Schau, et al. (1982a) obtained isotopic data from the Baker Lake region, 260 km west of the Daly Bay area. The Kramanitaur granulite-anorthosite complex, lithologically and structurally very similar to the Daly Bay Complex, has a U-Pb zircon age of 2573 $\pm^{28}_{5}$ Ma. It is in fault contact with the amphibolite facies Ingilik Point gneiss complex, that has a U-Pb zircon age of 2675 $\pm^{33}_{11}$ Ma. Whole-rock Rb-Sr data from the Ingilik Point gneiss complex give an isochron of 2069 ± 66 Ma, and an errorchron of 1908 Ma. K-Ar dates from the complex and younger gabbro stocks range from 1895 to 1982 Ma.

My interpretation is that the Daly Bay Complex may contain some rocks initially deposited or intruded in the Archean and that granulite facies metamorphism of these rocks occurred at approximately 2050 Ma. This metamorphism was terminated by tectonic emplacement into higher crustal levels which was complete at about 1950 Ma. Further uplift of the complex and surrounding rocks followed at about 1850 Ma.

ECONOMIC GEOLOGY

Gossans and rusty zones are present in the Outer Shear Zone, the Boundary Shear Zone and in quartzofeldspathic granulite. Most are in rocks with subeconomic graphite disseminations, although a few are associated with minor concentrations of pyrite or pyrrhotite.

The layered portions of the anorthosite bodies have generally been strongly sheared. Although oxide or sulphide layers are not known, their presence cannot be entirely discounted.

Minor amounts of chalcopyrite and pyrrhotite occur in unit Asq in a small gossan at 64°12'N, 89°50'W.

SPECULATION ON THE GEOLOGICAL HISTORY OF THE DALY BAY COMPLEX

Sedimentary rocks, protolith to parts of the Daly Bay Complex, were probably deposited in the late Archean. Minor amounts of sparsely preserved marble and quartzite in association with aluminous, probably pelitic, metasediments suggest a stable marine platform. These rocks were subsequently buried to a depth of at least 15 km. The cause of this burial is obscure, but could be related to the active tectonic environment that gave rise to Archean volcanic rocks in the Gibson-McQuoid lakes and Rankin Inlet map areas to the south (Reinhardt and Chandler, 1973; Davidson, 1970a, 1970b, 1972; Heywood, 1973). At some point the tectonic regime became sufficiently stable to permit mafic magmas to form layered gabbroic and anorthositic plutons.

At approximately 2050 Ma, these rocks began moving northward and upward to higher crustal levels. In the initial stages of this deformation, strain rates were sufficiently low with respect to ambient temperatures that continuous recrystallization of mineral grains took place at the same time the rocks were flowing to form large scale isoclinal folds. As temperatures dropped, the deformation mechanism changed and polygonization and disruption of mineral grains took place.

As rocks neared amphibolite facies conditions, strain became progressively localized in discrete zones. Gabbro dykes intruded but were themselves deformed as the rocks continued to undergo strain. At this time, what is now the Daly Bay Complex became surrounded by a ductile shear zone and behaved internally as a rigid plate.

At the leading edge of this plate, development of a series of shears and faults resulted in an interleaving of rocks originally part of the Daly Bay Complex with the rocks into which the complex was being emplaced. Granitic rocks intruded several of these shears. Continued cooling and concentration of strain led to the formation of several narrow cataclastic zones within the Outer Shear Zone. Deformation became negligible as the rocks reached greenschist facies conditions at about 1950 Ma.

Intrusion of dykes of the Mackenzie swarm at 1200 Ma was the last major event prior to erosional exhumation of the complex. Following the subsidence accompanying Ordovician and Silurian sedimentation in the Hudson Bay Basin, uplift, accompanied by north-northwest-trending faulting, brought the Daly Bay Complex to its present configuration.

The speculative history outlined above contains elements in common with the scenario for part of the Grenville Province outlined by Davidson (1984), which involves thrusting and stacking of large crustal blocks or domains, each of which is surrounded by a broad ductile shear band or bands. Metamorphic conditions preserved in the Daly Bay Complex lie between those of the Parry Sound Domain and Algonquin Park area (Schau et al, 1986). It is thus possible that the Daly Bay Complex is only part of a much larger region composed of imbricated slices of middle and lower crustal material.

Support for this hypothesis comes from examination of the regional gravity field. The Daly Bay Complex is associated with a local gravity anomaly superimposed on a regional high that extends eastward for 550 km from Baker Lake to Coats Island south of Southampton Island (Gibb and Halliday, 1975). The regional anomaly is associated with granulite gneiss and anorthosite at Baker Lake (Schau and Hulbert, 1977; Schau and Ashton, 1979, 1980; Schau et al., 1982b); with granulites along Chesterfield Inlet (Reinhardt and Chandler, 1973); a large gabbroic intrusion in the Armit Lake area (Heywood, 1967); the rocks described in this report; and meta-anorthosite on Coats Island and Walrus Island (Heywood and Sanford, 1976). The axis of the anomaly coincides with the Chesterfield Fault Zone of Heywood and Schau (1978).

To the southeast of the Chesterfield Fault Zone, prominent northeast-trending structures include the Tulemalu Fault Zone (Tella and Eade, in press), which may be extrapolated to intersect the Chesterfield Fault Zone at Baker Lake. Movement in both zones appears to have occurred from the late Archean through Early Proterozoic time. Although there are insufficient data to determine the relative timing and kinematic relationships of the two zones, is clear that the northeast trends of the southwestern Churchill Province do not extend through the Chesterfield Inlet area.

It is thus tempting to speculate that rocks along Chesterfield Inlet are remnants of Archean lower and middle crust emplaced into the upper crust by a major tectonic event during the Early Proterozoic. Documentation of the detailed kinematics of this event, and its relationship to other large scale crustal movements in the Churchill Province will be fascinating research topics in the years to come.

REFERENCES

Abbott, R.N.,Jr.
1982: A petrogenetic grid for medium and high grade metabasites; American Mineralogist, v. 67, p. 865-876.

Aranovich, L.Ya. and Podlesski, K.K.
1983: The cordierite-garnet-sillimanite-quartz equilibrium: experiments and applications; *in* Kinetics and Equilibrium in Mineral Reactions; Advances in Physical Geochemistry: v. 3., ed.S.K., Saxena. Springer Verlag, New York, p. 173-198.

Davidson, A.
1970a: Precambrian geology, Kaminak Lake map area, District of Keewatin; Geological Survey of Canada, Paper 69-51.
1970b: Eskimo Point and Dawson Inlet map areas (North Halves) District of Keewatin 55E and 55F (north parts); Geological Survey of Canada, Paper 70-27.
1972: The Churchill Province; *in* Variations in Tectonic Styles in Canada, ed. R.A. Price and R.J.W. Douglas; Geological Association of Canada, Special Paper Number 11, p. 381-433.
1984: Tectonic boundaries within the Grenville Province of the Canadian Shield; Journal of Geodynamics, v. 1, p. 433-444.

Fahrig, W.F and Jones, D.L.
1969: Paleomagnetic evidence for the extent of Mackenzie igneous events; Canadian Journal of Earth Sciences, v. 6, p. 679-688.

Froese, E. and Jen, L.S.
1979: A reaction grid for biotite-bearing mafic granulites; *in* Current Research, Part A, Geological Survey of Canada, Paper 79-1A, p. 83-85.

Ganguly, J. and Saxena, S.K.
1984: Mixing properties of aluminosilicate garnets: constraints from natural and experimental data, and applications to geothermobarometry; American Mineralogist, v. 69, p. 88-97.

Geological Survey of Canada
1973: Daly Bay; Geophysical Series (Aeromagnetic) Map 7606 G.

Gibb, R.A. and Halliday, D.W.
1975: Gravity measurements in northern District of Keewatin and parts of District of Mackenzie and District of Franklin, N.W.T.; Earth Physics Branch, Gravity Map Series 139-148.

Grant, J.E.
1985: Phase equilibria in partial melting of pelitic rocks; *in* Migmatites, Ed. J.R. Ashworth; Blackie, London., p. 86-144.

Heywood, W.W.
1966: Geological notes on Operation Wager, Northwest Territories; Geological Survey of Canada, Paper 66-10.
1967: Geological notes, northeastern District of Keewatin and southern Melville Peninsula, District of Franklin, Northwest Territories (parts of 46, 47, 56, 57); Geological Survey of Canada, Paper 66-40.
1973: Geology of Tavani map Area, District of Keewatin; Geological Survey of Canada, Paper 72-47.

Heywood, W.W. and Sanford, B.V.
1976: Geology of Southampton, Coats, and Mansel islands, District of Keewatin, Northwest Territories; Geological Survey of Canada, Memoir 382.

Heywood, W.W. and Schau, M.
1978: A subdivision of the northern Churchill Structural Province; *in* Current Research, Part A, Geological Survey of Canada, Paper 78-1A, p. 139-143.

Holdaway, M.J.
1971: Stability of andalusite and the aluminum silicate phase diagram; American Journal of Science, v. 271, p. 97-131.

Hutcheon, I., Froese, E., and Gordon, T.M.
1974: The assemblage quartz— sillimanite— garnet— cordierite as an indicator of metamorphic conditions in the Daly Bay Complex, N.W.T.; Contributions to Mineralogy and Petrology, v. 44, p. 29-34.

Low, A.P.
1906: Report on the Dominion Government Expedition to Hudson Bay and the Arctic Islands, on Board the DGS Neptune, 1903-1904; Ottawa, Government Printing Bureau.

Newton, R.N. and Haselton, R.T.
1981: Thermodynamics of the garnet— plagioclase— Al_2SiO_5— quartz geobarometer; *in* Thermodynamics of Minerals and Melts, Advances in Physical Geochemistry, v. 1, ed. R.C. Newton, A. Navrotsky and B.J. Wood; Springer Verlag, New York, p. 131-147.

Park, J.K.
1973: Paleomagnetism of metamorphic rocks of Daly Bay and Melville Peninsula, Churchill Structural Province; Canadian Journal of Earth Sciences, v. 10, p. 1079-1088.

Ramsay, J.G.
1967: Folding and Fracturing of Rocks; McGraw-Hill Book Company, New York, 568 p.

Reinhardt, E.W. and Chandler, F.W.
1973: Gibson-McQuoid lakes map-area, District of Keewatin; *in* Report of Activities, Part A, Geological Survey of Canada, Paper 73-1A, p. 162-165.

Robinson, J.L.
1968: Regional Geography; *in* Science, History, and Hudson Bay, v. 1., Ed. C.S. Beales; Energy Mines and Resources, Ottawa, p. 201-235.

Ruffman, A.
1969: Investigations of the crust in the Hudson Bay region; *in* Earth Science Symposium on Hudson Bay, Ed. P.J. Hood; Geological Survey of Canada, Paper 68-53, p. 307-311.

Schau, M.
1980: Zircon ages from a granulite-anorthosite complex and a layered gneiss complex northeast of Baker Lake, District of Keewatin; *in* Rubidium-Strontium and Uranium-Lead Isotopic Age Studies, Report 3, ed. W.D. Loveridge; *in* Current Research, Part C, Geological Survey of Canada, Paper 80-1C, p. 237-238.

Schau, M. and Ashton, K.E.
1979: Granulites and plutonic complexes northeast of Baker Lake, District of Keewatin; *in* Current Research, Part A, Geological Survey of Canada, Paper 79-1A, p. 311-316.
1980: Geological map of the granulite and anorthosite complex at the southeast end of Baker Lake, 56D1, 56C4, parts of 55M16 and 55N13; Geological Survey of Canda, Open File 712.

Schau, M. and Hulbert, L.
1977: Granulites, anorthosites and cover rocks northeast of Baker Lake, District of Keewatin; *in* Report of Activities, Part A, Geological Survey of Canada Paper 77-1A, p 399-407.

Schau, M. Davidson, A., and Carmichael, D.M.
1986: Granulites and granulites; Geological Association of Canada, Mineralogical Association of Canada, Canadian Geophysical Union, Joint Annual Meeting, Ottawa '86, Field Trip 6: Guidebook, 36 p.

Schau, M., Loveridge, W.D., and Stevens, R.D.
1982a: Updated Rb-Sr ages from the Ingilik Point Gneiss Complex, Baker Lake Region, District of Keewatin; *in* Current Research, Part C, Geological Survey of Canada, Paper 82-1c, p. 169-171.

Schau, M., Tremblay, F., and Christopher, A.
1982b: Geology of Baker Lake map area, District of Keewatin: a progress report; *in* Current Research, Part A, Geological Survey of Canada, Paper 82-1A, p. 143-150.

Tella, S. and Eade, K.E.
1986: The occurrence and possible tectonic significance of high pressure granulite fragments in the Tulemalu Fault Zone, District of Keewatin, N.W.T., Canada; Canadian Journal of Earth Sciences, v. 23, p. 1950-1962.